Adhesive Arachnoiditis
An Old Disease Re-Emerges in Modern Times

Adhesive Arachnoiditis

Copyright© 2019 Forest Tennant MD, MPH, DrPH

All rights reserved. This book or any portion thereof may not be reproduced or used in any manner whatsoever without the express written permission of the publisher except for the use of brief quotations in a book review or scholarly journal.

First Printing: 2019

ISBN 9780359822089

Arachnoiditis Research and Education Project
c/o Nancy Kriskovich
14 Hidden River Lane
Bigfork, MT 59911

Ordering Information:

Special discounts are available on quantity purchases by corporations, associations, educators, and others. For details, contact the publisher at the above listed address.

U.S. trade bookstores and wholesalers: Please contact Nancy Kriskovich Tel: (406)249-2002;
or email snkriskovich@gmail.com

All proceeds from the sale of this book will go to the Arachnoiditis Research and Education Project sponsored by the TENNANT FOUNDATION.
336 ½ S. GLENDORA AVENUE
WEST COVINA, CA 91790-3060
A 501(c)(3) Non-profit Organization

About the Authors

Dr. Forest Tennant Biography 2019

Dr. Forest Tennant has had a long, distinguished career in addiction and pain medicine. He was recently honored with a "Life-Time Achievement Award" by Pain Week and "50 Years of Service" by the journal, "Practical Pain Management" where he served as their medical editor for 12 years. Dr. Tennant began his career in addiction and pain medicine as a US Army medical officer during the Vietnam War Era at which time he helped develop the Army's drug testing, education and treatment programs. After the Vietnam War, he became a Public Health Fellow assigned to the UCLA School of Public Health where he obtained his Masters' and Doctorate in Public Health (MPH, DrPH). Since 1975 to the present day, Dr. Tennant has produced a steady flow of research and publications in addiction and pain medicine. During this time his public notoriety first emerged when he served as an expert witness in the criminal trials of the physicians of Howard Hughes and Elvis Presley. Later he developed the drugs of abuse and anabolic steroid testing programs for the Los Angeles Dodgers and the National Football League. In very recent years Dr. Tennant has chosen to research hormonal testing and treatment of intractable pain and the merging pain problems of adhesive arachnoiditis and genetic collagen disorders. He and his wife Miriam of 52 years reside and work in West Covina, California and Wichita, Kansas.

Ingrid Hollis Biography 2019

Ingrid Hollis began her career as a classically trained artist. This life path was put on hold by the necessity of becoming a family caregiver. This experience, along with a life altering motor vehicle accident in 1996 and the discovery that she has hypermobile Ehlers Danlos syndrome, (hEDS), a genetic collagen disorder that also runs in her family, changed the direction of her life. This knowledge set her on a course to search for answers, medical help and cures.

A great curiosity for research was ignited through this direct experience and fueled the desire to help others in similar situations. She has studied traditional and non-traditional medicine from many remarkable physicians and healers from throughout the world. Her hope has been to glean the wisdom from the many different healing systems and find ways to help and improve the lives of patients and their families who suffer from rare diseases throughout the world. Ingrid resides in Colorado with her retired husband and son. She has two lovely children and two grandchildren to whom her work is dedicated.

She heads the "Publication Team" for the Arachnoiditis Research and Education Project of the Tennant Foundation.

An Old Disease Re-Emerges in Modern Times

Forest Tennant

Dedications

Dr. Antonio Aldrete
Dr. Margaret Aranda
Dr. Lloyd Costello
Dr. Scott Guess
Dr. Ryle Holder
Dr. Adam Hy
Dr. Terri Lewis
Dr. Caren Pederson
Dr. M. J. Porcelli
Dr. Sara Smith
Terri Anderson
Pat Anson
Lynn Ashcraft
Steve Ariens
Louise Carbonneau
Donna Corley
Anne Fuqua
Heather Grace
Matthew Guffan

Gabriella Gutierrez
Ted Knopf
Melanie Lamb
Kate Lamport
Red Lawhern
Denise Molohan
Kristen Odgen
Amy Patridge
Rhonda Posey
Gary Snook
Vance Snyder
Debra Vallier
Kris Walters

In memory of...
Jennifer Adams
Claudine Goze-Weber

Adhesive Arachnoiditis

ADHESIVE ARACHNOIDITIS

AN OLD DISEASE RE-EMERGES IN MODERN TIMES

Forest Tennant MD, MPH, DrPH
Ingrid Hollis

Adhesive Arachnoiditis

CONTENTS

1. PRECIS/ABSTRACT---11
2. INTRODUCTION – ARACHNOIDITIS IN MODERN TIMES-----------13
3. HISTORY OF ARACHNOIDITIS---17
4. THE REASONS WHY ADHESIVE ARACHNOIDITIS IS RE-EMERGING--21
5. MODERN DAY CAUSES --27
6. NATURAL COURSE OF THE DISORDER-------------------------------------35
7. SYMPTOMS---39
8. PHYSICAL SIGNS--43
9. LABORATORY TESTS---45
10. DIAGNOSIS OF ADHESIVE ARACHNOIDITIS-------------------------47
11. TREATMENT--49
12. PREVENTION---55
13. SUMMARY---57
14. APPENDICES---59

    APPENDIX-I MODERN TIMES PROMOTERS OF ADHESIVE ARACHNOIDITIS ---61

    APPENDIX-II A SELF-SCREENING TEST FOR LUMBAR-SACRAL ADHESIVE ARACHNOIDITIS --63

    APPENDIX-III ADHESIVE ARACHNOIDITIS CATEGORIZATION----65

    APPENDIX- IV PROCEDURE FOR THE DIAGNOSIS OF LUMBAR-SACRAL ADHESIVE ARACHNOIDITIS --------------------------------67

Adhesive Arachnoiditis

PRECIS/ABSTRACT

Arachnoiditis is an inflammation of the middle layer of the meninges, which is the protective covering that surrounds the brain, spinal cord, and cauda equina nerve roots. The meninges (spinal canal covering) is connective tissue composed mainly of collagen and elastin. Arachnoiditis has been known since the 19th century when its major cause was infections, mainly tuberculosis and syphilis. During the 20th century the major cause of arachnoiditis was the use of insoluble oil-based dyes that were injected into spinal fluid to enhance x-ray views, called myelograms, for diagnostic purposes. In this century adhesive arachnoiditis is the condition of most concern. It is caused by the adherence or "gluing" of the cauda equina nerve roots to the arachnoid spinal canal lining by adhesions. The major causes of adhesive arachnoiditis are anatomic abnormalities and from genetic collagen disorders, autoimmune diseases, and traumatic injuries. Recently, the scientific understandings of both neuroinflammation and neuroregeneration has led to the development of treatment protocols for adhesive arachnoiditis. While formerly viewed as an untreatable disease, this reputation is no longer true.

INTRODUCTION – ADHESIVE ARACHNOIDITIS IN MODERN TIMES

Arachnoiditis is simply defined as inflammation of the arachnoid membrane which is the middle layer of the meninges, the protective covering tissue that envelopes and encases the brain and spinal cord.[1] The outer layer of this covering is the dura mater and the inner most layer is called the pia mater. Since the arachnoid layer covers the entire brain and spinal cord, inflammation of the arachnoid can occur in many locations from a variety of causes. Trauma, infection, hemorrhage, or a cancerous growth involving the arachnoid layer may cause inflammation, and will be labeled according to its cause and/or anatomic location.[2] Terms such as cerebral, neoplastic, or thoracic arachnoiditis have been used. Although the arachnoid layer is the one layer most involved with the inflammatory process, the other two layers, the pia mater and dura, may also become inflamed.[3] The most

Adhesive Arachnoiditis

prevalent forms of arachnoiditis occur in the lumbar-sacral (lower back) and cervical (neck) regions of the spinal cord.[4,5]

Thanks to modern diagnostic technology of contrast magnetic resonance imaging (MRI), the term lumbar-sacral adhesive arachnoiditis has come into usage, and this form of arachnoiditis is the major concern in modern times.[6-8] This term is applied when a lumbar-sacral MRI shows that some cauda equina nerve roots have clumped together and formed adhesions to the arachnoid layer of the spinal canal covering (meninges). This condition entraps, inflames, and renders the affected lumbar-sacral nerve roots impaired and dysfunctional. Neurologic impairments and severe, devastating pain may result, since the nerve roots connect to and regulate the nerves of the gastrointestinal tract, urinary system, sex organs, and lower extremities. All or some of the nerve roots may become impaired and dysfunctional when nerve conduction is blocked by inflammatory damage to the nerves. Besides direct neurologic impairments and severe pain from trapped and damaged nerves, adhesive arachnoiditis may produce spinal fluid flow obstruction and chronic leakage or "seepage" through the dural layer into the soft tissues that surround the spinal column. Since lumbar-sacral adhesive arachnoiditis is the most prevalent form of arachnoiditis that causes severe disability, suffering, and pain, this book only refers to lumbar-sacral adhesive arachnoiditis unless specifically noted.

Adhesive Arachnoiditis

HISTORY OF ARACHNOIDITIS

The word "arachnoid" derives from the arachnoid membrane's fragile spiderweb like appearance.[2] Arachnoiditis as a distinct disease was first recognized in the mid 1800's. The common causes of arachnoiditis then were the infections of tuberculosis and syphilis. The famous French neurologist, Jean Martin Charcot, recorded a description of the disorder in 1869.[9] Dr. Thomas Addison, the British physician who discovered adrenal failure, published his findings on 11 autopsied patients in 1855.[10] Two of these cases had severe pain and atrophied adrenals with calcium deposits and fluid around the arachnoid layer. This suggested that long-standing, end stage arachnoiditis was the likely cause of severe pain and consequent adrenal failure. The exact year the disorder was named and credited is uncertain, but the 1873 "Comprehensive Medical Dictionary" published by J.B. Lippincott & Co. included this definition: "Arachnitis: A faulty term, denoting inflammation of the arachnoid membrane."[1] The same dictionary simply defined arachnoiditis as, "inflammation of the arachnoid membrane."[1]

The first recorded attempt to treat arachnoiditis was probably in 1781 when Dr. John Fothergill, a famous British physician, treated a patient with severe back and sciatic pain along with other symptoms compatible with arachnoiditis.[11] Opioids failed to relieve the pain, but a mercury concoction resulted in a positive result. In 1899 the first edition of the Merck Manual listed drugs and measures for spinal meningitis with notations that the recommendations were for chronic and tubercular meningitis.[12] No less than 30 recommendations were made for treatment, including opium and water baths which are still used today. In 1909 the neurosurgeon, Sir Victor Horsley, recognized inflammation of the meninges to be from infection or hemorrhage, and he coined the term "chronic spinal meningitis" which established the notion that inflammation involved all layers of the meninges.[3] Harvey, in 1926, determined that adhesion formation that connects nerve roots, spinal cord, or brain to the meninges to be a major cause of pain and disability.[13] He established the concept of a serious pathological partnership involving the nerve roots and the arachnoid layer.

The major cause of arachnoiditis changed as infectious disease treatment agents were developed in the 1900's for syphilis and tuberculosis. A new cause, however, originated in about 1930, when poorly soluble oil dyes were infused into the spinal canal to enhance x-ray visualization for diagnostic purposes.[2,3,13] These x-rays were called myelograms. A small percentage of patients who received these dyes developed arachnoiditis. Magnetic Resonance Imaging (MRI) technology was developed about 1987. MRI's progressively replaced oil-based dyes and consequently, for a time, arachnoiditis seemed to almost totally disappear. New developments, however, have caused this disorder to reappear in modern times.[6-8]

Beginning around the year 2000, an increasing lifespan, obesity, and sedentary lifestyle, all began to contribute to a huge increase in degenerative disorders of the spine. Accidental traumas such as whiplash due to an auto accident or repetitive stress injuries due to motions from performing a job or playing sports can cause progressive spinal deterioration, over time, as well. These factors, along with highly technical medical and surgical interventions, have all contributed to the emergence of adhesive arachnoiditis in today's world.

THE REASONS WHY ADHESIVE ARACHNOIDITIS IS RE-EMERGING

On the surface, it appears that the reason for re-emergence of adhesive arachnoiditis is primarily the result of patients with chronic, degenerative spine conditions that undergo epidural injections or surgery rather than pursue less risky therapies. This explanation requires a more in-depth analysis primarily directed at the reasons for such a high prevalence of chronic, degenerative spine conditions. First, people are living longer, and longevity, per se, seems to be associated with osteoporosis, osteoarthritis, and intervertebral disc atrophy with protrusion into the spinal canal. Adhesive arachnoiditis is now being recognized in patients in their 70's and 80's.

Factors that are known to foster lumbar spine degenerative condition are well known and include obesity, poor posture, bucket seats, lack of exercise, and a sedentary lifestyle. The latter is obvious in modern living as each day we all encounter people constantly sitting and watching television, along with their computer, lap top or new "know-it-all" cell phone. "Tech neck" is a new term for cervical spine issues arising from this poor posture practice.

The rage to place bucket seats in cars and airplanes hasn't helped prevent lumbar-sacral disease. While bucket seats and seat belts are clearly lifesaving in an accident, they distort the anatomy of the lower spine. Bucket seats put pressure on the intervertebral discs and cause the cauda equina nerve roots to compress on each other. Good old-fashioned padded, bench seats should be the staple at home or work where high impact, collisions can't happen. Many jobs besides those that require sitting in front of a computer screen may entail repetitive trauma to the lumbar-sacral spine. Besides living longer, humanity has been blessed with nutritional, safety, and medical measures that allow them to live through childhood infections and genetic disorders, but also leave them at risk of developing adhesive arachnoiditis in adulthood. It appears as though genetic diseases may be increasing. In prior generations, poor immunity increased deaths of many infants and children who contracted an infectious disease. Today vaccines and antibiotics now allow some persons to become adults. Unfortunately, some of these genetic and immune compromising disorders will leave them at risk for developing adhesive arachnoiditis in adulthood. Included in this list are sickle cell disease, Ehlers-Danlos Syndrome, dwarfism, and some autoimmune diseases such as psoriasis. The amazing emergency medical and other trauma systems of today have saved thousands of persons after severe motor vehicle or industrial accidents. The person is saved, but, is left with spinal injury the possibility of developing lumbar-sacral adhesive arachnoiditis.

An interesting phenomenon is the discovery that the infection from Lyme disease bacteria can also cause adhesive arachnoiditis.

Syphilis was a major infection that caused adhesive arachnoiditis in the 19th century. Lyme disease bacteria and syphilis are in the same biologic classification known as spirochetes.

Modern medicine has given us the epidural anesthetic injection for childbirth labor and diagnostic spinal taps. While thousands of these procedures are done with no complication, each one carries a low, but very real risk of adhesive arachnoiditis.

In summary, no one would want to give up childhood antibiotics and vaccines, computers, delicious foods, safe-driving cars, cell phones, or our cutting-edge medical emergency systems. All these marvels of modern times, however, require awareness that they may contribute to adhesive arachnoiditis. Simple measures and awareness of risk factors can, fortunately, prevent adhesive arachnoiditis while one still enjoys and benefits from the miracle of modern medical treatment.

Adhesive Arachnoiditis

MODERN DAY CAUSES

In the 21st century there are 4 basic causes of adhesive arachnoiditis: (1) anatomic disorders of spinal structure; (2) genetic collagen disorders; (3) trauma, and (4) autoimmune disorders.[6,7,8] Adhesive arachnoiditis may be initiated by either damage to the arachnoid lining or to the nerve roots in the cauda equina. Regardless of which site the damage starts, eventually adhesions will adhere or glue nerve roots and arachnoid lining together which will cause severe pain and impairment. The spinal cord ends at about the top of the lumbar spine and turns into about a dozen nerve roots. They are free floating in the spinal fluid canal ("pipe") and are collectively known as the cauda equina. At some point along the lumbar and sacral spine nerve roots individually exit to connect and carry nerve impulses from the brain to the feet, legs, bladder, sex organs, intestine, and stomach.

Adhesive Arachnoiditis

Any anatomic alteration that causes the spinal canal to narrow, bend, or otherwise distort in the lumbar-sacral region of the spine has the potential to eventually cause adhesive arachnoiditis. The spinal canal, which contains the cerebral spinal fluid, must always remain open and free from obstruction or blockage. Spinal fluid is made in the brain and it recreates itself about 4 times a day. It leaves the brain to enter the spinal canal.[14] Functions of cerebral spinal fluid are to carry nutrients, remove contaminants such as inflammatory byproducts, and constantly bathe and lubricate the nerve roots of the cauda equina so they don't rub on each other. Blockage of flow may cause irritation, neuroinflammation, scarring, and adhesion formation.[14-16] The spinal canal which carries the spinal fluid must constantly remain open and clear so there is no fluid flow obstruction or compression of nerve roots.[16] It is for this reason that the human body is designed to move, stand and walk every few minutes and turn often during sleep. There are several spinal conditions that can interfere with spinal fluid flow and cause obstruction and compression, which can irritate the nerve roots in the cauda equina. Spinal/vertebral arthritis, osteoporosis, vertebral collapse, and intervertebral disc herniation may be age or accident related. Kyphoscoliosis, spondylolisthesis, and rheumatoid spondylitis are genetically caused. The most common anatomical structural condition which causes adhesive arachnoiditis is chronic intervertebral disc protrusions (herniations). The protruding discs may narrow or compress the spinal canal, either constantly or intermittently, and force nerve roots in the cauda equina to rub together and eventually form inflammatory clumps. These inflammatory clumps may subsequently develop adhesions which adhere or glue to the arachnoid lining.

Once inflammation develops in the cauda equina nerve roots, any trauma, including medical interventions or toxic substance in the spinal fluid may accelerate the formation of nerve root clumping and adhesions.

A recently recognized pathologic process that may cause adhesive arachnoiditis is genetic, autoimmune, and viral disorders that directly cause degeneration and deterioration of the arachnoid and other layers of the spinal canal covering. The genetic conditions are called "collagen tissue disorders" of which the best known are the Ehlers-Danlos/Marfan syndromes.[17-20] Collagen dissolution is the pathologic mechanism which deteriorates the very fragile tissues of the spinal cord and canal such as the pia mater and arachnoid linings[17-20] Besides adhesive arachnoiditis, other spinal conditions may result from this process and include perineural (Tarlov) cysts, syringomyelia, tethered cord, and Chiari disorders.[17-20] Patients with genetic collagen disorders are at high risk and more prone to experience injuries and adhesive arachnoiditis from intrathecal, epidural, and surgical procedures due to softer, inherently fragile connective tissues. Once the arachnoid layer is damaged and inflamed by a genetic or autoimmune disorder, it may "capture" and "glue" nerve roots to itself by adhesion formation.

Autoimmune disorders can also attack spinal tissue and cause inflammation in the arachnoid layer.[21-23] Autoimmune disorders that are associated with adhesive arachnoiditis include systemic lupus erythematosus, rheumatoid arthritis, and psoriasis.

Trauma, both from accidents, and from medical interventions in the form of medical injections and surgery, may cause adhesive arachnoiditis.[24-27] Severe, acute trauma from accidents that affect the lumbar-sacral spine, including motor vehicle, work-related, or falls may incite the development of inflammation in either the arachnoid layer or nerve roots. Once initiated, the inflammation may proceed to adhesion formation. Unfortunately, there are now well-documented cases where a single spinal tap or epidural injection for therapeutic or anesthetic purposes has caused adhesive arachnoiditis. In some cases, there is no known underlying anatomic structural, genetic, or autoimmune condition to predispose to the development of adhesive arachnoiditis.

At this time, the precise mechanism for the rare development of adhesive arachnoiditis after a medically justified spinal tap or epidural injection is unclear. A spinal tap is clearly a traumatic procedure and it is possible that a toxic contaminant or infectious agent could enter the spinal fluid in the spinal canal through the needle track. Epidural injections of anesthetics, corticosteroids, or other medications are almost always safe when administered under fluoroscopy. Rarely is there trauma to the dural and arachnoid linings. The epidural space is, however, not totally separate from the subdural and/or subarachnoid space.[24-27] Many thousands of villi, or granules, are present on the arachnoid layer that form connecting channels between the epidural veins and spinal fluid.[25] These channels allow any chemical, such as an anesthetic, corticosteroid, contrast agent, or solvent injected into the epidural space to potentially enter the spinal fluid.

Although the majority of diagnosed, clinical arachnoiditis cases are in the lumbar-sacral region, some arachnoiditis cases occur in the cervical (neck) region. The diagnosis in these cases is almost entirely clinical as there is no equivalent to the cauda equina nerve root complex to be observed on MRI. Cervical neck arachnoiditis, in the author's experience, is almost entirely in patients with degenerative spine conditions and stenosis who have also had surgery and/or multiple epidural injections. Cervical arachnoiditis is manifested by severe pain on either forward or backward extension of the neck, but not both. There may be arm and hand neurological deficiency signs. MRI images suggest a thickened arachnoid layer with clearly visible spinal fluid flow obstruction.

NATURAL COURSE OF THE DISORDER

Without treatment, arachnoiditis has a somewhat unpredictable natural course. Some cases may resolve naturally, and do so completely with no evidence that it existed.[2,4,5] Once adhesive arachnoiditis is established, however, it may be classified as mild, moderate, severe, or catastrophic. A mild case is one in which pain is intermittent and there is no impairment of bladder function or extremity range of motion. A moderate case is one with constant, but easily controlled pain, and is accompanied by some impairment of bladder function and weakness of the lower extremities. A severe case will require daily pain control medication as the pain is severe, constant, and intractable. There is impairment of bladder function, inability to sit or stand in one position for more than a few minutes, along with weakness or paraparesis (partial paralysis) of the lower extremities. A catastrophic case will have all the elements of a severe case plus partial paralysis of the lower extremities, cognitive impairment, severe fatigue, bed or couch bound existence, and will demonstrate a need for assistance to carry out activities of daily living. When left untreated, catastrophic cases of adhesive arachnoiditis will leave the patient with a shortened life span and autoimmunity. Terminal events are usually related to overwhelming infections or cardiac-adrenal failure.[33] A major goal of modern treatment is to prevent the disease from progressing from a mild or moderate form to a catastrophic state.

The seriousness of adhesive arachnoiditis is directly related to the intensity of glial cell activation and neuroinflammation in the nerve roots of the cauda equina.[28-31] Serum testing of neuroinflammatory markers has revealed that neuroinflammation may fluctuate from high intensity to being relatively innocuous or inactive.[32] The natural, untreated course of adhesive arachnoiditis is, therefore, unpredictable.

In addition to spontaneous resolution, some cases appear, for unknown reasons, to have active neuroinflammation that stops or "dies off" but leaves behind permanent neurologic impairments and pain. Most cases that come to the attention of today's physicians tend to be progressive with intermittent periods of remission and activity.

The catastrophic stage of adhesive arachnoiditis may be disappearing likely due to medical and physical measures that do not cure or eliminate the disorder but do seem to prevent the most catastrophic and devastating complications such as lower extremity paralysis, adrenal failure, severe immune deficiency, and early, life-ending infection and sepsis.[33] To support this possibility, MRI's of today's patients rarely show calcification (arachnoiditis ossificans) which was typical and well-recorded in the last century[2]. In summary, modern day treatment measures, whether specifically directed at adhesive arachnoiditis or other spine disorders, seem to have reduced the occurrence and severity of complications and the devastating progression to a catastrophic state.

SYMPTOMS

The typical symptom profile of adhesive arachnoiditis is constant lumbar pain with a variety of neurologic manifestations.[7,8] Pain intensity may change with movement and positional changes between sitting, reclining, or standing. Persons with adhesive arachnoiditis may achieve comfort while standing or lying down, or the situation may be just the opposite, either sitting or lying down may increase pain. In addition to positional pain relief, other symptoms experienced by the majority of adhesive arachnoiditis patients include the sensations of water dripping and/or bugs crawling on the legs. Burning sensation on the bottom of the feet is also common. Some urination difficulty is usually present. It may involve hesitancy in starting flow of urine or difficulty stopping (neurogenic bladder). Dribbling between urinations is also common. Many patients complain of blurred vision, headache, and dizziness, which are believed to be due to spinal fluid flow obstruction. Nerve root clumps and scar tissue present within the spinal canal act as a "dam" obstructing constant spinal fluid flow. Spinal fluid may leak, or more appropriately "seep", through the arachnoid and dura layers of the spinal canal covering into the soft tissues surrounding the spinal canal. Cerebral spinal fluid is extremely toxic and caustic to soft tissue, so considerable pain and damage will develop. Soft tissue including muscles, nerves, and fascia may degenerate, scar, and shrink after coming in contact with the leaking fluid.

Adhesive arachnoiditis, like other brain and spinal cord diseases, can create a systemic or generalized autoimmune disorder with immunodeficiency.[21-23] This is believed to occur when inflammatory by-products and/or brain or spinal cord tissue reach the general circulation.[22,23] Autoimmune manifestations include arthritis of joints, muscle pain, thyroiditis, and small fiber neuropathies. Some adhesive arachnoiditis patients are first diagnosed with an autoimmune disorder such as systemic lupus erythematosus or Hashimotos/Thyroiditis. Some are diagnosed with fibromyalgia.

Since nerve roots of the cauda equina have multiple connections to various internal organs, a variety of symptoms may occur depending upon the anatomical location of the inflamed and entrapped nerve roots. These may include nerves to the stomach and intestine, including the rectum and anus. Food sensitivity, nausea, vomiting, constipation, diarrhea, urinary and fecal incontinence may all occur in different patients. Sex organ, bladder and bowel function may be adversely affected. In severe cases incontinence and impotence may occur. Breathing symptoms, including shortness of breath may occur. The heart may race (tachycardia). Leg and feet impairments are extremely

Adhesive Arachnoiditis

common including foot drops, unsteady gait, and weakness in the legs and feet. Partial paralysis is relatively common and even total paralysis from the waist down has been known to occur.[7,8]

PHYSICAL SIGNS

The major physical signs in an adhesive arachnoiditis patient are to be found in the lower extremities and back. Physical signs in the legs include weakness, instability, poor balance, and abnormal reflexes. Pain may occur with extension or stretching of the legs. Partial paralysis of the lower legs may be present. Adhesive arachnoiditis patients usually have more pain on one side of the body than the other. Consequently, the patient will often tend to favor one side, constantly leaning in a position that lowers their pain. Over time, this attempt to find comfort and relief produces asymmetrical muscle groups in the back with observable areas of muscle hypertrophy and atrophy.[8]

If paraspinal muscles and soft tissues have been bathed in chronic cerebral spinal fluid seepage, they may scar and shrink. Consequently, patients may not be able to fully extend their arms and legs. Chronic cerebral spinal fluid seepage may cause considerable shrinkage of the soft tissues between the skin and spine resulting in an indentation of midline tissue along the lower spine.

LABORATORY TESTS

Currently, there is no specific laboratory test that identifies an adhesive arachnoiditis patient. Since adhesive arachnoiditis is a neuroinflammatory disease, some by-products of inflammation known as "markers", are often elevated. These include the erythrocyte sedimentation rate (ESR); C-reactive protein (CRP), myeloperoxidase, and cytokines.[32] If any of these markers are elevated, treatment goals should include suppressing neuroinflammation and lowering the affected marker into its normal serum range. It is emphasized that an absence of elevated inflammatory markers does not necessarily mean that neuroinflammation is either controlled or absent.

Some adrenal and gonadal hormones such as cortisol, testosterone, and pregnenolone may drop in the serum due to the severe stress and pain of adhesive arachnoiditis.[34] Serum hormone levels that are low are regarded as indicators that the disease process is not well-controlled.

DIAGNOSIS OF ADHESIVE ARACHNOIDITIS

The diagnosis of adhesive arachnoiditis requires 4 elements:

1) History of an inciting event or disease

2) Typical symptoms

3) Abnormal physical signs

4) MRI findings. Specific evidence of adhesive arachnoiditis is usually recognized on a contrast MRI which is one that uses injected dye or a high-resolution 3 Tesla (without dye) imaging to contrast the spinal fluid from the spinal cord, nerve roots and arachnoid covering.

The necessary evidence to make a diagnosis of adhesive arachnoiditis is;

1) The presence of nerve root clumping

2) Adhesions which attach the clumps to the arachnoid layer of the spinal cord covering.

While other indicators such as nerve root displacement, enlargement, and asymmetry may be present on MRI imaging, these findings, by themselves, are insufficient to warrant a diagnosis of adhesive arachnoiditis.[4-6]

A major issue is that the classic appearance of nerve root clumping and adhesion formation may not be visible for several months after an inciting event such as a spinal tap or epidural injection. Also, patients may have symptoms, physical signs, and laboratory test abnormalities that indicate adhesive arachnoiditis is present but not show MRI evidence of adhesive arachnoiditis. In these cases, therapeutic trials with medications used to treat adhesive arachnoiditis are warranted.

TREATMENT

Until recently, adhesive arachnoiditis has been called "untreatable" and "hopeless."[2,5] Two major scientific discoveries have made it possible to develop a "first generation" medical process or protocol to treat it. The first discovery is that neuroinflammation is caused by activation of cells in the brain and spinal cord called "glia."[28-31] Pain, injury, infection, or exposure to foreign chemicals or metals (such as those that enter the spinal fluid with medical interventions and surgical procedures) activate glia cells which produce neuroinflammation. Since these discoveries, some medicinal agents and hormones have been identified that will suppress glial cell activation and neuroinflammation.[29,31]

The second discovery is that the brain and spinal cord make certain hormones called neurohormones whose primary functions are to suppress neuroinflammation and/or regenerate damaged nerve cells.[35-41] Some are called "neurosteroids" since they contain a steroidal, chemical structure.[35,36] These include pregnenolone, allopregnanolone, progesterone, dehydroepiandrosterone, and estradiol. Administration of some of these neurosteroids has been well-demonstrated to control neuroinflammation and promote neuroregeneration in laboratory animals.[38,40,41] The administration of some neurosteroids and their chemical analogues are now being used to treat adhesive arachnoiditis.[8]

The treatment process of adhesive arachnoiditis consists of two basic elements:
1) **Medication;**
2) **Physiologic measures.**

Medications consist of 3 components:
1) **suppressors of neuroinflammation (examples: ketorolac, methylprednisolone)**

2) **Neuroregeneration (re-growth), (examples: pregnenolone, nandrolone)**

3) **Pain control, (examples: low dose naltrexone, gabapentin, opioids).**

Physiologic measures are targeted at maximizing spinal fluid flow, prevention of scarring and shrinkage of nerve roots, muscles, and other potentially affected cells that may cause neurologic impairments and pain.[15,16] Basic physiologic measures include daily walks, gentle stretching of the extremities, water soaking, deep breathing, and light weightlifting.

Pain control for adhesive arachnoiditis is symptomatic and basically follows that of standard pain care. Unfortunately, the pain of adhesive arachnoiditis may rival, or exceed that of metastatic bone cancer and of necessity require last-resort, symptomatic measures such as implanted electrical stimulators and high dose opioids including those administered by injection, suppository and implanted intrathecal pumps. Many new pain treatments are being investigated for severe intractable pain similar to that caused by adhesive arachnoiditis. Some, such as intravenous infusions of lidocaine, vitamin C, and ketamine are reportedly providing extended periods of pain relief. While pain control is purely symptomatic, the physiologic measures and medical agents to suppress neuroinflammation and promote neuroregeneration are implemented to permanently bring about some disease resolution, diminution of symptoms and impairments, and enhanced quality of life.

Readers of this book need to be clearly advised that the treatment process for adhesive arachnoiditis is new; a "first generation" effort at effective treatment and which must be considered elementary. The current protocol will undoubtedly be changing as new discoveries are made. While the three components of medical treatment;

1) Suppression of neuroinflammation

2) Promotion of neuroregeneration

3) Pain control

are enduring concepts, specific treatment agents, including the examples listed here, should all be considered a "first generation effort." Improvements will be developed when additional clinical studies, including controlled drug trials, are launched.

Adhesive Arachnoiditis

PREVENTION

The finding that many cases of adhesive arachnoiditis are preceded by degenerative or structural abnormalities of the spine suggest that primary prevention of adhesive arachnoiditis is possible. Chronic spinal degenerative conditions are well known to be associated with sedentary lifestyles, obesity, and lack of exercise. It is also well known that back pain and injuries may too often be subjected to high technology and invasive medical procedures before non-invasive measures are fully explored. While adhesive arachnoiditis is seldom or rarely the result of a single spinal tap or epidural injection, persons who have a degenerative, structural, or a genetic collagen disorder should be well-aware of the risk of adhesive arachnoiditis with these procedures.[24-27] Consequently, anyone considering these procedures is well-advised to consider non-invasive measures for back and spine conditions, and seek second opinions before embarking on invasive medical procedures including surgery.

Persons who develop lumbar pain with leg and bladder dysfunction immediately following a medical procedure including spinal tap, epidural anesthesia, or surgery are at high risk of developing adhesive arachnoiditis. If symptoms of adhesive arachnoiditis such as severe back and leg pain with urinary dysfunction occur after a medical procedure, in order to prevent adhesive arachnoiditis, it is recommended that the most potent anti-neuroinflammatory agents such as ketorolac and methylprednisolone be administered on an emergency basis.

SUMMARY

Multiple factors of modern day living have caused adhesive arachnoiditis to re-emerge after being almost invisible for a generation. Consequently, adhesive arachnoiditis is now a condition that is present in the population of every modern community.

The diagnosis of adhesive arachnoiditis is made when a person has a typical history, symptoms, physical findings, and an MRI that show cauda equina nerve root clumping with adhesions to the arachnoid lining.

In the past, with no treatment available, this disorder caused excruciating pain, paralysis, immobility, incontinence, adrenal failure, and early death. Hence, its reputation as both "hopeless and untreatable." Thanks to multiple recent scientific discoveries, a medical treatment process or protocol has been developed consisting of 3 medical components.

3 MEDICAL COMPONENTS

(1) Suppression of neuroinflammation

(2) Promotion of neuroregeneration

(3) Pain control

Although it is in its first generation, this process is spreading in medical practice and has been receiving a most positive, albeit anecdotal, response. The process will undoubtedly improve with controlled studies and more widespread clinical experience.

APPENDICES

Adhesive Arachnoiditis

APPENDIX I

MODERN TIMES PROMOTERS OF ADHESIVE ARACHNOIDITIS

- ✓ Sedentary (sitting) lifestyle*
- ✓ Poor posture
- ✓ Non-supportive footwear
- ✓ Bucket seats
- ✓ Risk-taking sports/activities
- ✓ Obesity
- ✓ Lack of exercise and walking
- ✓ Risky spinal injections and surgery
- ✓ Longer lives and aging
- ✓ Emergency life saving measures following accidents**
- ✓ Childhood survival with antibiotics and vaccines***

*Refers to the long hours of sitting in front of television and computers
**Life may be saved but injury left behind
***Immune-deficient infants may survive and develop tissue degeneration disorders in adulthood

Adhesive Arachnoiditis

APPENDIX II

A SELF-SCREENING TEST FOR LUMBAR-SACRAL ADHESIVE ARACHNOIDITIS

	QUESTIONS	Y	N
1	In addition to chronic pain, do you ever experience sharp, stabbing pains in your lower back when you twist, turn or bend?		
2	Do you ever experience bizarre skin sensations such as crawling insects or water dripping down one or both legs?		
3	Do you ever have burning, tingling, or a sensation of walking on broken glass in your feet and/or toes?		
4	Does your pain become worse while standing, sitting and/or walking?		
5	Do you have leg weakness and/or pain that radiates down one or both legs?		
6	Do you experience any bladder dysfunction such as dribbling, or difficulty when starting or stopping urination?		
7	Do you sometimes have a headache along with blurred vision?		

If you answered yes to four or more of these seven questions, you most likely have adhesive arachnoiditis or some other neuroinflammatory disease of the nerve roots in your lumbar or sacral spine. Your physicians need to be informed of the results of this screening test as you need to obtain a confirmatory diagnosis.

Adhesive Arachnoiditis

APPENDIX III

ADHESIVE ARACHNOIDITIS CATEGORIZATION

MILD		
	A	Full range of motion
	B	No back indentation or contracture
	C	Normal inflammatory markers
	D	No bladder impairment
	E	No MRI evidence of spinal fluid leakage or obstruction
	F	No hormone abnormalities
	G	Can sit and stand in one position for 10 minutes
	H	Intermittent pain
MODERATE		
	A	Full range of motion and walks without assistance
	B	Mild to zero lower extremity weakness
	C	Normal inflammatory markers
	D	Some bladder hesitancy, urgency, dripping
	E	No MRI or physical evidence of spinal fluid leakage
	F	Mild constant pain but no need for sleeping medication
	G	Can sit and stand in one position for 10 minutes
SEVERE		
	A	Some range of motion impairment and may need assistance (cane or other) to ambulate
	B	Weakness in lower extremities with neurologic symptoms (e.g. burning feet, bugs crawling, jerking or other
	C	Elevated inflammatory markers and/or hormone abnormalities
	D	Bladder impairment symptoms of hesitancy, urgency, or incontinence
	E	MRI and/or physical evidence of chronic spinal fluid leakage and/or flow obstruction
	F	Constant pain that impairs sleep
	G	Can't sit and stand in one position for 10 minutes
CATASTROPHIC		
	A	Requires assistance with activities of daily living (dressing, toiletry, eating, etc.)
	B	Significant lower extremity impairment (e.g. walker, wheelchair, braces)
	C	Bladder impairment of hesitancy, urgency, or incontinence
	D	Mental deficiencies such as memory loss or reading ability
	E	MRI and physical evidence of chronic spinal fluid obstruction and leakage
	F	Elevated inflammatory markers and hormone abnormalities
	G	Constant pain that impairs sleep
	H	Can't sit or stand in one position for 10 minutes

A person does not need to have all items in each category as listed here and may have some variations in each category.

Adhesive Arachnoiditis

APPENDIX IV

PROCEDURE FOR THE DIAGNOSIS OF LUMBAR-SACRAL ADHESIVE ARACHNOIDITIS

I. <u>HISTORY – COMMON PREDISPOSING CONDITIONS</u>
   1. Pre-existing spine condition: herniated discs, Kyphoscoliosis, arthritis, osteoporosis, stenosis, spondylolisthesis
   2. Genetic/Inherited Disorders: Ehlers-Danlos/Hypermobile Syndromes, Marfan syndrome, Tarlov cysts, rheumatoid spondylitis, Chiari malformation.
   3. Inciting or Triggering Event: trauma, surgery, spinal-epidural injection, electrocution, infection, myelogram, caudal block, spinal cord tumor.

II. <u>COMMON SYMPTOMS</u>
   1. Back and buttock pain that radiates to legs
   2. Bladder dysfunction
   3. Bizarre skin sensation (crawling insects, water dripping)
   4. Burning feet
   5. Leg weakness
   6. Positional pain (worse or improved on standing/sitting)
   7. Blurred vision
   8. Leg pains, cold sensations
   9. Loss of feeling in extremities

III. PHYSICAL FINDINGS – NON-SPECIFIC/COMMON
1. Leg weakness – one side is common
2. Pain – straight leg raising
3. Loss touch/vibration sensation in foot, ankles, cold to the touch
4. Restriction of range-of-motion in arms and/or legs
5. Decreased reflexes – unilateral
6. Pain on pressure over lower lumbar-sacral area
7. Asymmetry of back musculature
8. Indentation of mid-back/spine area

IV. LABORATORY – SERUM ABNORMALITIES
Interpretation: Low level – likely excess pain and inflammation

    1. Hormone panel: cortisol, pregnenolone, DHEA, progesterone, estradiol, testosterone

Interpretation: If any elevated – likely excess of neuroinflammation

    2. Inflammatory markers: erythrocyte sedimentation rate (ESR), C-reactive (CRP), Cytokines

NOTE: Normal blood tests do not rule out arachnoiditis or the presence of neuroinflammation.

V. MAGNETIC RESONANCE IMAGING (MRI)
SOME CAUDA EQUINA NERVE ROOTS MUST BE CLUMPED AND ADHERED BY ADHESIONS TO THE SPINAL CANAL LINING

References

1. Thomas J. Arachnitis and Arachnoiditis. *Comprehensive Medical Dictionary*. Philadelphia, J.B. Lippincott & Co. 1873;p57.
2. Aldrete JA. History and evaluation of arachnoiditis: the evidence revealed. *Insurgentes Centro 51-A. Col San Rafael, Mexico* 2010; p3-14.
3. Horsley VAH. Chronic spinal meningitis: its differential diagnosis and surgical treatment. *Br J Med* 1909;1:513-517.
4. Quiles M, Marchiselo PJ, Tsairis P. Lumbar adhesive arachnoiditis: etiologic and pathologic aspects. *Spine* 1978;3:45-50.
5. Burton C. Lumbosacral arachnoiditis. *Spine* 1978;3:24-30.
6. Jackson A, Isherwood I. Does degenerative disease of the lumbar spine cause arachnoiditis? A magnetic resonance study and review of the literature. *Brit J Radiology* 1994;67:840-847.
7. Aldrete JA. Suspecting and diagnosing arachnoiditis. *Pract Pain Mgt* 2006;16(5):74-87.
8. Tennant F. Arachnoiditis diagnosis and treatment. *Pract Pain Mgt* 2016;14(7):63-69.
9. Charcot JM, Joffrey A. Deax cas d'atrophic musculaire progressive avec lesions de a substance gris et des faisceaux anterolateraux de la moelle spinaire. *Arch de Physiologic* 1869;2:354-358.
10. Addison T. The constitutional and local effects of disease of the supra-renal capsules. *London:Samuel Highley*;1855.
11. Elliott J. Of the cure of sciatica in a complete collection of the medical and philosophical works of John Fothergill. *Pater-Nofter-Row*, London;1781:p355-363.
12. Meningitis, Cerebral, Spinal, and Tubercular in Merck's 1899 Manual of the Materia Medica: A Ready Reference Pocket Book for the Practicing Physician. *Merck & Co. New York*, 1899;p146.
13. Harvey SC. Meningeal adhesions and their significance. *Interstate Post Grad Med, North America Prac* 1926;2:27-31.
14. Damkier HH, Brown PD, Praetorius J. Cerebrospinal fluid secretion by the choroid plexus. *Physiol Rev* 2013;93(4):1847-92.

15. Kiiski H, Aanismaa R, Tenhunen J, et al. Healthy human cerebral spinal fluid promotes glial differentiation of hESC-derived neural cells while retaining spontaneous activity in existing neuronal networks. *Biology Open* 2013;2(6):605-612.
16. Whendon JM, Glassey D. Cerebrospinal fluid stasis and its clinical significance. *Altern Ther Health Med* 2009;15(3):54-60.
17. Henderson FC, Austin C, Benzel E, et al. Neurological and spinal manifestations of the Ehlers-Danlos Syndromes. *Amer J Men Gen* 2017;175C:195-211.
18. Castori M, Voermann NC. Neurological manifestations of "Ehlers-Danlos Syndromes. *Iran J of Neurol* 2014;13:190-208.
19. Sevesto S, Merli P, Ruggier M, et al. Ehlers-Danlos Syndrome and neurological features: a review. *Childs Neuro Syst* 2011;27:365-371.
20. Schievink WI, Gordon OK, Tourje J, et al. Connective tissue disorders with spontaneous spinal cerebrospinal fluid leaks and intracranial hypotension: a prospective study. *Neurosurgery* 2004;54(1):65-71.
21. Wang L, Wang FS, Gershwin ME. Human autoimmune disease: a comprehensive update. *J Intern Med* 2015;278:369-395.
22. Zweiman B, Levinson AI. Immunologic aspects of neurological and neuromuscular diseases. *JAMA* 1992;268(20):2918-2922.
23. Meisel C, Schwab JM, Prass K, Meisel A, Dirnage U. Central nervous system injury-induced immune deficiency syndrome. *Nat Rev Neurosci* 2005;6(10):775-786.
24. Epstein NE. The risks of epidural and transforminal steroid injections in the spine: commentary and a comprehensive review of the literature. *Surg Neurol* 2013;4(supe2):574-593.
25. Eisenberg E, Goldman R, Shclag-Eisenberg D, Grinfeld A. Adhesive arachnoiditis following lumbar epidural steroid injections: a report of two cases and review of literature. *J Pain Research* 2019;12:513-518.
26. Nelson DA. Dangers from methylprednisolone acetate therapy by intraspinal injection. *Arch Neurol* 1988;45(7):804-806.
27. Kitson MC, Kostopanagiotau G, Alimeric K, et al. Histopathological alterations after single epidural injection of rapivacaine, methylprednisolone acetate, or contrast material in swine. *Cardiovasc Intervert Radioil* 2011;34(6):1288-1295.

28. O'Callgan JP, SriranT, Miller DB. Defining "neuroinflammation." *Ann NY Acad Sci* 2008;1139:318-330.
29. Tsuda M. Microglia in the spinal cord and neuropathic pain. *J Diabetes Investig* 2016;7(1):17-26.
30. Loggia MI, Chunde DB, Oluwaseum A, et al. Evidence for brain glial activation in chronic pain patients. *Brain* 2015;138:604-615.
31. Mika J. Modulation of microglia can attenuate neuropathic pain symptoms and enhance morphine effectiveness. *Pharmacol Rep* 2008;60:297-300.
32. Bilello J, Tennant F. Patterns of chronic inflammation in extensively treated patients with arachnoiditis and chronic intractable pain. *Postgrad Med* 2016;92(17):1-5.
33. Javidi E, Magnus T. Autoimmunity after ischemic stroke and brain injury. *Fron in Immunol* 2019;10:1-12.
34. Tennant F. The physiologic effects of pain on the endocrine system. *Pain Ther*. 2013 Dec;2(2):75-86.
35. Compagnone NA, Mellon SH. Neurosteroids: biosynthesis and function of these novel neuromodulators. *Front Neuroendocrinol* 2000;21:1-56.
36. Jones KJ. Gonadal steroids and neuronal regeneration: a therapeutic role. *Adv Neurol* 1993;59:227-240.
37. Joels M, DeKloet E. Control of neuronal excitability by corticosteroid hormones. *Trends Neurosci* 1992; 15:25-30.
38. Guth L, Zhang Z, Roberts E. Key role for pregnenolone in combination therapy that promotes recovery after spinal cord injury. *Proc Natl Acad Sci* 1994;91:12308-12312.
39. Kilts JD, Tupler LA, Keefe FJ, et al. Neurosteroids and self-reported pain in veterans who served in the military after September 11,2001. *Pain Med* 2010;10:1469-1476.
40. He J, Evans CO, Hoffman SW, et al. Progesterone and allopregnanolone reduce inflammatory cytokines after traumatic brain injury. *Exp Neuro* 2004;189:404-412.
41. Patil AA. The effect of human chorionic gonadotropin (HCG) on restoration of the spinal cord: A preliminary report. *Int Surg* 1990;75(1):54-57.

Adhesive Arachnoiditis

NOTES

CPSIA information can be obtained
at www.ICGtesting.com
Printed in the USA
BVHW032101300320
576429BV00003B/9